Nick's Notes™

NOTE #1: These Sativa-dominant *Lamb's Bread* clones, said to have been Bob Marley's favorite strain, have just begun to flower – while they take considerably longer to mature and are more difficult to manage due to their greater overall height and girth (easily reaching 5 or so ft. indoors and 8 or so ft. out), I prefer their more energetic, 'soaring' effects, especially during the active, working daylight hours (whereas I smoke more Indica-dominant strains at night)

My First Half-Dozen Indoor Medical Marijuana Growing Lessons

An Outline on the Most Essential Information for Indoor Marijuana Cultivators

(Written by a California Medical Marijuana Licensee for his Fellow Licensees)

Nick's Notes ™

My First Half-Dozen Indoor Medical Marijuana Growing Lessons

1. **Organic fungicide and pesticide mixes can prevent many of the most common problems for indoor growers, but *don't forget to heavily dilute all such preventive mixes before using them to dip or spray your plants* - it is very easy to harm or even kill the plant in its very sensitive early stages with this otherwise worthwhile preventative measure, as most pest and fungal attacks can be prevented by applying such solutions to your plant(s) before placing them in your garden(s)** - plants are most vulnerable as young vegetating seedlings/clones because they've yet to develop their natural defenses, such as the 'cystolith hairs' that are about the size of and interspersed between the resin glands (the "trichomes" or 'THC crystals') but without their luster, that produce a substance that gums-up the mouths of their would-be predators - keep in mind that, due to their natural grow environments, Indica's are more naturally resistant to pests and Sativa's are more naturally resistant to fungal attacks - when applying protective sprays/baths to plants, I suggest that beginners start by using just one brand of preventative solution at a time and by mixing no more than one total *teaspoon* [depending, of course, on the concentration of the solution(s) - *always* read *all* information provided on product bottles *before* applying *anything* to your plant(s)] of the fungicide/pesticide mixture(s) into a relatively small but deep, basin-like container filled with water - while combining two or more types of solutions/mixtures is common practice among experienced growers, be *very* careful what you combine as some of the oils in these solutions don't mix well together and can be lethal to foliage when combined, especially in greater volumes and higher concentrations - although there are both far too many effective preventative solutions and potential lethal combinations of those solutions and their constituent oils to mention herein, from my limited experience I've used and found both Ed *Rosenthal's* *Zero Tolerance* and *Organic Laboratories'* *Organocide* not just very effective examples of such products but an excellent example of a pairing that can very easily be lethal when mixed into the same bath/spray - *always* start with the recommended dosage levels printed on all product bottles and then adjust from there as your experiences and observations dictate - regarding the actual pests most likely to contest the indoor grower's healthy harvest, while you will need other resource(s) for a more comprehensive list of plant pests and the signs of their infestations, look-out for plant-killing mites especially: while *spiders* are generally an ally to indoor and outdoor growers alike, consuming many plant-eating insects, spider *mites* are one of the growers' greatest enemies - mostly invisible to the naked eye, if permitted entrance into your garden area(s) these infinitesimal creatures will colonize your garden(s) and then literally begin to suck the life from the underside of your plants' leaves - spider mites first show signs of their infestation via the presence of a whitish to yellowish speckling pattern on the tops of leaves (that is generally more diversified, better defined and less 'powdery looking' than the visible signs of the onset of the "powdery mildew" fungus noted in #3 herein) paired with fine,

hair-like webs lining plant leaves that, if permitted to progress, will thicken and eventually surround and suffocate plants all the way up to the pinnacle of their primary cola - the fine webs produced by spider mites can be *very* tricky to pick-up on, especially in its early stages and especially without *constant* vigilance - make their webs more clearly identifiable by misting your plants with water which, in the worsening stages of their infestation, will reveal not just their webbing but tiny flecks in those webs that are actually the mites' gestating larvae (egg sacks) - the crop-killing little bastards are reproducing in your garden(s) ! - without the presence of their natural predators, like the lady bug, the prevention of such potentially-disastrous infestations must be accomplished by (a) utilizing dips/sprays, by (b) assuring that all plants in your garden are *constantly* supplied with a steady dose of fan-generated, artificial, colony-disturbing 'winds,' by (c) maintaining clean growing area(s) and, in a related matter, by (d) making absolute certain *never* to bring plants, containers, tools or anything else that's been sitting outdoors *into* your protected indoor growing environment(s) - with the application of the preventative bath(s)/spray(s), while I don't do this unless there are signs of an infestation or fungal attack, it *may* be worth your while to apply the bath/spray a second time, likely via spray considering the plants will be substantially larger now, before repotting the plants for the flowering room/area, assuming a two-part "perpetual harvest" type operation is being employed (which is *highly* recommended – read on for an explanation) - in terms of spray technique, considering that spider mites and many other pests lodge and consume carbohydrates from the undersides of the plant's leaves, I recommend that, once they've been identified in an established garden and spraying is deemed necessary, that you mix a half to one *teaspoon* of one of your preventative oil mixtures into a standard-sized spray bottle and *spray from the bottom up* so that the pest colonies usually established under the colonized plants' leaves are adversely affected, if not eliminated and *always spray your plants just before the lamps in their grow space(s) are switched off for the day* (read on for an explanation) - *keep experimenting* (in *all* matters related herein) until you discover your ideal levels of dilution and your favored brand(s) and types of preventative mixtures - make sure to keep your preventative dip/spray solution constantly mixed, or "agitated," throughout the dip/spray process to avoid allowing the constituent oils to settle into over-concentrated globules, risking the formation of dead 'burn spots' on leaves – after the dip/spray solution is applied to your clone(s)/seedling(s), it is *highly*-recommended that you pick-up every doused clone/seedling and, holding it in its small pot/container in one hand, carefully grasp the baby Buddha plant near the base of its stalk with the other hand and shake it vigorously (but *not* violently) for a good 20 seconds or so until you can no longer feel the droplets of your preventative bath/spray flying-off of the plant's vulnerable early-growth leaves – *always* allow newly-dipped or sprayed plants at least 6 to 8 hours to *completely* dry-out before placing them under your lamp(s), as the oils in such anti-fungal and anti-pest mixtures will produce a chemical reaction in the leaves when mixed with *direct* light that will burn and kill-off plant leaves and even destroy entire plants if they are placed under the lamp(s) before the oils in the utilized mixture(s) are given the chance to *entirely* penetrate the plant's sensitive early-growth leaves - I personally killed many young clones making this mistake - place dipped/sprayed plants in a dark, well-ventilated area during this drying period - you can also soak a towel or washcloth as soon as you've dipped/sprayed your last clone/seedling and hang it in the same space as your wet clone(s)/seedling(s) – as soon as you can use that towel/cloth to comfortably and fully dry yourself off after a shower or shave your new clones/seedlings are likely ready for insertion in the garden(s) – keep plants 2-3 feet or so from the light source(s) for the first 2-3 days after your dip/spray application if using non-high-intensity discharge lamps to vegetate, like those using compact fluorescent bulbs, and more like 3-4 feet or so if using an HID lamp (likely a metal halide) - after this settling-in period your plants can go as close as 6 inches to non-HID's (like CFL's - see #4) and no closer than a foot to most HID's - regarding the choice of whether or not to grow from clone or from seeds, many accomplished growers attest to the productive benefits of growing from the more genetically-rich seed, despite the added time and cost involved, as opposed to growing from clones, which are often a copy of a copy ad indefinite

2. **Over-fertilization and over-watering are two of the most common and interrelated mistakes made by beginning indoor growers - assuming you've set-up a 2-plus-stage cultivation project where you're 'perpetually harvesting' every month or so (see #4 herein) and aren't starting-out by putting your clone(s)/seedling(s) in 3 or 5 gallon container(s), it is *very* easy to give just-developing plants way too much water and nutrients to clones/seedlings before they've had the chance to establish strong roots, as the excess water and fertilizer has much less space to saturate in smaller containers - therefore make sure to heavily dilute your nutrient mixtures before feeding in the beginning while plants are building their 'root balls'** - graduate feeding of both water and nutrients as the plants begin to fill-out the vegetation space - closets are often used for the vegetation stage of development as such closed-off areas (commonly referred to as "vegetation closets" by indoor growers) are good at trapping moisture (vegetating plants prefer a more humid environment than do flowering plants – more on this in #3) and containing the often less-concentrated light given-off by more economical but less-powerful non-HID lamps like compact fluorescents ("CFL's" - #4 covers most lighting-related notes) that are commonly used during the vegetation stage usually lasting for about 2 weeks to a month - the 2 most common signs of over-fertilization are oddly and unevenly-shaped leaves and/or leaves that are a concentrated dark forest green in color and possibly starting to splotch or streak with bright yellow – when your plants start to show such signs you can, depending upon how clear the signals are of there being a problem present: (a) feed them a heavily-diluted ratio of nutrient-mixture-to-water for a few days to a week, (b) give the overfed plants exclusively water for a few days to a week or, (c) combine a few days to a week of heavily-diluted feedings or just watering *and* a "leeching" (aka a "draining" or "scrubbing") - take the plant(s) to the sink or tub and run their containers through with water for about twenty seconds or so to allow most of the over-concentration of nutrients to leech out of the medium held within that container - to prevent such buildups of nutrients from forming within your growing container(s) and the medium(s) within them in the first place is to give your plants solely water once or twice a week or about a third of the time you feed/water them, even if there aren't any visible signs of over-fertilization (it is, as with #1, a *preventative* measure) - healthy roots appear a vibrant white, while drowning roots are generally brown and noticeably unhealthy in appearance - the roots of over-

watered plants can't literally can't breathe and gradually begin to drown, resulting in the wilting, drooping leaves of a dying plant - the issue of *under*-fertilization, on the other hand, is rare, especially among beginners that tend to assume more of everything must be better - lack of sufficient nutrients manifests in plants much the opposite of the way over-fertilization manifests, producing leaves that are a pale green in color to the point of bordering on a dull yellowish hue and generally appearing less healthy than well-fed foliage - when watering plants in their established gardens don't allow the excess water/solution that has seeped all the way through container(s) to sit in the receptacles beneath the container(s) for too long, as the water covering the bottom holes prevents ideal root-oxidization levels (it hinders proper "breathing" and draining through the container's bottom holes) - regarding the 'catch-all' receptacles that you need to put under your plant(s) to catch excess solution that drains through the plant's container(s) before it floods your grow area floor(s), use something, like plastic saucers, that fit that bottom of your container(s) well enough to hold a good deal of water without overflowing but not so large as to prevent your being able to huddle your plants together to economically make use of your light source(s) - assuming you haven't established an automated hydroponic system that recycles the excess drain-through water and nutrients for you, keep a turkey baster or similar moisture-extraction tool handy so that the excess fluid that consistently drains through your growing container(s) can be removed when necessary - keep in mind that as new imports to your vegetation area, baby clones/seedlings will only require about a 15th to a 10th of a gallon of water per day to start to maintain strong, consistent growth, while fully-flowering Sativa-dominant strains (like the *Lamb's Bread* clones planted in 5 gallon containers pictured on the cover herein) sitting under 1,000W High Pressure Sodium (HPS) lamp(s) in 5 gallon containers can require as much as one full gallon *apiece* every day - in terms of the actual nutritional regimen and hydroponic setup to apply to your marijuana plants, there are *far* too many reputable brands, strategies and methods to adequately explore herein - thus far I've used only the water-by-hand-to-drained-catch-all container system known in hydroponics as the "Drain-to-Waste Method" because the excess 'caught' water and nutrients are dumped down the drain or lost to evaporation, or are 'wasted' rather than recycled (though you can always use a baster or similar liquid-extraction tool to suck-up and immediately reapply or store the excess water/nutrients - just remember to *re-stir* if you extract and reuse - read on for more on this topic) - this newcomer grower has thus far used the 2-part *FloraNova* nutrient regimen series by *General Hydroponics* as well as their 3-part series consisting of *FloraGro*, *FloraMicro* and *FloraBloom*, both with good results *when properly applied* – regarding watering by hand, provide water from your jug until you can just start to notice the excess solution draining into the catch-all basins or saucers below, then cease watering - no matter your undertaking, every endeavor requires the right set of tools - make use of *at least* a set of measuring spoons and a solid mixing stick or utensil (I use a 16" piece of dried bamboo I found to mix the nutrient solution with water in my 2 gallon jug – it works perfectly) when feeding, and follow the instructions on the container(s) when mixing the concentrated nutrient fluid with clean, *drinkable* water (the better it tastes to you the better it tastes to your plants - if the water supply at your grow site tastes 'off,' it may be worth investing in a filtering water jug or a reverse-osmosis machine to filter your water before applying it your system or hand-watering it directly into the grow container(s) at the base of your plant(s) - add some or all of the water you plan to use for every to-be-mixed feeding solution to your watering jug or container *before* mixing-in your nutrients to prevent any of the solution from settling on the bottom and coming-out as half-dissolved sludge when the container is tipped upside down while running dry (I sludged one baby clone to death this way when I first started growing) - *keep the growing medium(s) in your planted container(s) consistently 'moist' as opposed to 'wet' or 'soaked' as, again, too much water will literally drown your plants* - in terms of the actual <u>growing mediums</u> to employ in your indoor growing project, there are numerous that will work well, including many possibilities that aren't popular or even commercially-viable - a substance or mix of substances need only possess certain select characteristics, like the proper level of potential-hydrogen, or "PH," and sufficient capacities for water retention and the permitting of root-penetration in order to pass for a proper "growing medium"), thus far I like the water-retaining and draining characteristics of coconut coir, made from the fibrous outer husk of the coconut fruit, while many of the more experienced growers seem to prefer the cleanliness of using trays with inserted blocks of rockwool or the simplicity and reusability of clay pellets, to name but a few of the most popular types (advanced 'aeroponic' systems use only moisturized air for a 'growing medium) - *as a recurring theme herein, only the chief teacher in all things, <u>experience</u>, can award you with ideal watering/feeding volumes and schedules and the best growing medium for you* - avoid watering leaves directly since, similar to but less extreme as with the preventative bath/spray noted in the first part of this outline, permitting droplets of *any* liquid/solution to sit on the tops of plant leaves for too long can be detrimental to those leaves when that sitting moisture is mixed with the effects of direct light, often resulting in dead spots (consider that in the majority of the globe's natural ecosystems this rainbow-friendly weather of both rain and *direct* sunlight simultaneously falling upon the plant life is rare, especially in the Indica's natural ecosystems) - the consideration of additives or 'boosters' to your base nutrient system is another undertaking too considerable to relate in detail herein, other than to note that it's a realm worth exploring and that even the beginning grower should utilize, at the bare minimum, <u>some brand of root-benefiting mixture containing any of that class of helpful microbial lifeforms that literally bind-to and create a symbiotic relationship with the roots of plants by which they considerably expand upon the growth rate and, therefore, upon the nutrient and water-intake-capacity of the supplied plants' root system</u> in return for siphoning-off a negligible measure of water and nutrients for themselves - numerous brands and types of these mixtures are, like everything else the indoor grower could ever need, available at any of the ever-more-omnipresent hydroponics shops spreading throughout California and even at some licensed dispensaries (such as *Organicann* in my area) - while, again, there are several different types of these root-benefiting microbes made commercially available, the one most often used by marijuana cultivators and the only one I've thus far utilized is called "<u>Mycorrhizae</u>" and is available in soluble and insoluble form from many reputable companies – combine water with the suggested quantity of Mycorrhizae plus an optional "root booster" nutrient mix like *Bio Root* by *General Organics* and apply the mixture to your clones/seedlings before planting them in the vegetation area - apply a similar but stronger solution a 2nd time when repotting the same plants in their 3 or 5 gallon container(s) just before their insertion in the flowering area in another 2 to 4 weeks - <u>look for the plant(s) to show their roots, which should be white if healthy, through the bottom holes of your containers as a sign of when they need to be repotted to avoid stagnating their development</u> - while 3 gallon container(s) are all you'll need for the Indica plants' condensed indoor lifecycles, the substantially-larger Sativa-dominant and especially the pure Sativa strains won't fulfill their potential, even indoors, without being planted in the larger 5 gallon container(s) - while growing large quantities of quality bud indoors is definitely doable, keep in mind that the indoor grower consciously trades the productive *quantity* potential of the outdoor grower for the prospect of growing a final product that's much higher in *quality* (in medicinal potency and corresponding value per gram/ounce) – <u>when you pot/repot your plants</u> it is suggested that the bottom inch or so of your growing container(s) be layered with gravel/stones or something similar, as the pressure differential created assists drainage, and that the top inch or so be filled with something like peat moss or wood chips, as both tactics will promote consistent watering, temperature and moisture-retention levels within the growing medium - although the top layer of 'mulch' will prevent some of the water/nutrients from actually making it to the roots, requiring that you provide more of each than would otherwise be required, and although the bottom drainage-assistance layer will keep your plant(s) from completely filling their container(s) with their root systems and will thus minimally restrict growth and production, the previously mentioned benefits of both of these potting tactics are generally considered worth the extra costs involved - <u>regarding repotting technique</u>, first water the to-be-planted medium with a diluted water/nutrient mixture and compact the medium (assuming it's something like coconut coir and not something non-compactable like a cube of rockwool or clay pellets, simply press down on the top of the medium within the container), then use a small spade or your hands to dig a hole roughly the same size and shape of the smaller container from which the plant is being removed [as previously mentioned the plant(s) being repotted should have filled-out its/their container(s) with its/their root system(s) as a secondary sign of needing to be repotted] - *<u>when removing plant(s) from smaller container(s) make sure to disturb the root system as little as possible</u>*, then place the plant(s) being repotted in their newly-dug hole(s) in the larger container(s) - pack extra medium around the base of the stem and add your top layer of mulch (chips, moss etc.) and compact around the base of the plant's stem again for added stability and root-to-soil contact - give plant(s) more water than usual during this vulnerable stage of root re-expansion - <u>final important water note</u>: anticipate the day of harvest and perform a <u>final flushing</u> of the to-be-harvested female plant(s) to assure that the majority of the bad-tasting chemicals derived from the watered-in nutrient mixture(s) (assuming you're not simply using soil or some other purely organic feeding medium and growing method) in the container(s) and plant(s) are washed away by giving your almost fully-ripened females nothing but water (optionally mixed with a specially-made 'flushing solution') for at least their final week

3. **<u>Proper ventilation and circulation is vital to the successful growing of a fruitful harvest – without sufficient air movement through your grow area(s) your plants will not only grow at a retarded rate, if they live long at all (as available oxygen</u>**

and carbon dioxide, both vital to plant growth and overall health, will be at a minimum), but they will be _much_ more prone to both infestations of plant-eating-organisms such as the plant-strangling spider-mite that thrives in an environment with poor circulation (as natural or fan-produced 'winds' disturb their under-leaf feeding and colonization), and will invite communities of fungi that thrive in poorly ventilated areas with relatively high levels of moisture content, or "humidity" - one of the most common types of fungi to invade indoor gardens is <u>powdery mildew</u>, which manifests as white splotches on the leaves and looks as if the leaves have literally been sprinkled with baking soda, except it doesn't dust-off like powder because it's alive - _never forget that fungi thrive in poorly-ventilated areas that hold or 'trap' a relatively high degree of moisture_ - the amount of relative moisture present in air is called _humidity_ and is tracked with a "hygrometer" that is usually coupled to a thermometer - moisture will be present in higher than normal concentrations in the air of every grow area due to the fact that grow areas contain regularly-watered and transpiring ('sweating') plants, and because the excess water fed to and _not_ taken-in by the roots of those plants is being naturally evaporated into the air and will generally remain there unless vented out of the room/area - <u>_always_ make certain that every grow area has a way to _both_ draw-in fresh air from the outside as well as to expel or 'push-out' the already-utilized-air</u> into which the plants have released the by-products of their development (their "waste") as the consistent presence of oxygen and CO2-thin, stagnant air in any growing environment will, as in the "weakest chain in the link" or "slowest cog in the clock" analogies, bring the development of otherwise sufficiently-supplied plants to a near grinding-halt - to avoid the stunted growth produced by low-quality, stale air, <u>_don't allow grow room air to be trapped_</u>: assure that all of your primary ventilation-based fans and ducting or built-in vents are aligned in such a way as to suck air into your grow area(s), past your garden(s) and _all the way through_ the growing area(s) - make sure your secondary, circulation-based adjustable fans are situated such that they blow their air into and work _with_ rather than against the current of nutrient-rich air that should be continually being brought-in from the outside by your primary, usually more powerful ventilation fans before pushing that air across the tops of your plant(s) (as, again, the upper 'canopy' of the garden is generally the area most at risk of overheating and harboring spider mites and other plant-eating organisms (refer to #1 for more info on preventing pest infestations) - after being pushed over, around and through your garden(s) your circulating grow room air should, ideally, be blown towards and sucked up and into ducting (_since heat naturally rises, grow room air heated by large HID lamps will naturally want to rise <u>up and out</u> of the grow room/area_) or else should be pushed across the room towards one or more of the ventilation-based 'outtake fan(s)' that should be positioned to push stale air out the opposite end of the room from where the 'intake fan(s)' is/are pulling fresh air into the area - _when you can't physically alter the space you're growing in, as with leased apartments such as mine, instead of ducting your stale, used-up air up into the attic and then out through the roof, you'll have to push the air utilized by your garden(s) into your living quarters, so make sure that if you have housemate(s) they are OK with the ongoing odors that will be produced by such a less-than-ideal setup_ - to promote proper quality-air-uptake around your garden make sure to leave a good 2 to 3 feet around the perimeter of every set of plants huddled together under each lamp in order to maximize light utilization, as this open perimeter is not only necessary for proper circulation but for the proper access to your plants that will be vital for their ongoing feeding, watering, rotating and general upkeep - when estimating the proper size of this perimeter gap keep in mind that your plants, *especially* your flowering plants and especially if they're Sativa's or Sativa-dominant, will not only grow in height but will considerably expand in width as well, so make proper allowances if you don't want to have to re-hang your usually heavy, cumbersome HID equipment later - vegetating clones/seedlings focused on root development prefer a more humid environment than do maturing, flowering females – according to Cervantes (see "Author's Acknowledgments" towards the end herein), vegetating plants prefer a relative humidity range of 60 to 70 percent, while flowering plants prefer a range somewhere between 40 to 60 percent – with humidity being a relative measure of the amount of moisture present in the air in any area at any one time, and with the moisture in the air of any indoor growing environment being primarily derived from the water/nutrient solution being provided to the plants inhabiting that environment, the bottom, or "floor," of these ranges can provide a quick, rough indicator for when your plants should to be watered again – for example, when relative humidity hits about 40 percent in my flowering room I can be fairly certain that the plants in the room are relatively thirsty and could use some water – Sativa's can tolerate a higher degree of relative humidity compared to Indica's because they're tropical plants that evolved in ecosystems characterized by their containing copious quantities of both heat and moisture (with humidity, the higher the temperature the more moisture the air can hold) - regarding carbon dioxide (CO2), a critical element in plant development, one air-quality trick for providing more CO2 to your plants *without* having to invest in expensive CO2-producing-machinery is to pair your indoor growing project with a brewing hobby – the reason, other than the fun involved in producing and consuming your own libations, of course, is that CO2-production is an abundant by-product of the fermentation process by which yeast microbes will convert *any* form of sugar into alcohol, especially in warm environments (heat activates or 'wakes up' the yeast) - a less involved, time-consuming and cost-intensive option here is to simply brew something you don't plan on drinking, as it is the actual *process* of brewing that matters to your plant(s) and *not* the final result (the brew itself): simply continue to dump brewers yeast, any form(s) of sugar (cane sugar, brown sugar, honey, left-over candy etc.) and hot water (*not* boiling water, as the excessive heat could kill the yeast and negate your efforts) into a basin to be left in your grow area(s) - whatever your brewing tactic, if any, brew in close proximity to your garden(s) for optimal benefit - another excellent reason for keeping your grow space(s) well-ventilated is odor detection - trapping odor so that others can't smell it may *seem* like a good idea at first, but not only is trapped air low-quality, stagnant air, but upon its inevitable escape from your flowering area(s) (vegetating plants generally don't emanate strong odors like flowering plants) those odors will have concentrated to the point of being *much* more noticeable than if that air was properly blown through and vented out of your grow space(s) – the use of carbon filters, to which growers commonly affix inline fans that suck-in and deodorize grow room air before forcing that air up attached ducting and pushing it through the area's HID hood(s) (often with the assistance of inline fan(s) affixed to built-in chambers on the HID hoods themselves) before that air is expelled from the growing area(s) is the most common multi-pronged strategy used by experienced growers to control the moisture, heat and odors produced by grow areas in order to avoid run-ins with ambitious law enforcement officials and disapproving, nosy neighbors, of which there always seems to be at least one within close proximity of most residential areas - I also read in a 2011 issue of *High Times Magazine* of a growers' use of a crock pot to cook meals whenever his plants were approaching maturation and the estimated day of harvest, as the first type of 'pot' provided a handy method by which to conceal the distinct odor(s) emanating from the second

4. **Sufficiently-concentrated and spectrum-appropriate light source(s) represent the most commonly lacking component and, thus, the most prevalent growth-restricting factor holding-back the insufficiently-planned, researched and/or grower-experience-constrained indoor medical marijuana growing operation** – considering indoor light sources must compete with the sun, which not only provides an ideal degree of total illumination, which can be measured in its output of "lumens," but the color or "spectrum" of the light provided by the sun to the plant life in any ecosystem has evolved in partnership with those plants to create an ideal environment for them first to *spring* into life (they like light closer to the blue end of the color spectrum in the early part of their "vegetating" lives), then begin to thrive in the peaking summer of their lives, when somewhere between the blue and red ends of the light spectrum, or "white light," is ideal [that portion of the light spectrum that can be utilized by plants is called the "photosynthetic active region," which is why in hydroponics you sometimes hear the expression "PAR Watts," which refers to that portion of the electricity your lamp(s) convert into light that your plants can actually utilize for their development] and, finally, look to the completion of their life-cycles during the fall when the average angle of the sun has receded to the degree in the sky that, as it enters the atmosphere, emits light closer to the red end of the spectrum, somewhat similar to the spectrum of light produced by the sun's rays as it begins to set during the daily hours of dusk - indoors everything provided by Mother Nature must be simulated, of course, which is why the general consensus among experienced growers is that it is preferential to first put new clones/seedlings in a dedicated vegetation area (I use a coat closet set near the entry to my apartment for my "vegetation closet") that is illuminated with a whitish-blue light such as that produced by the metal-halide (MH) high-intensity-discharge (HID) lamps preferred by experienced growers due to their greater overall productive capacities, or the less-wasteful *non*-HID compact fluorescent (CFL) or light-emitting-diode (LED) lamps and, once those plants have fully rooted into their smaller containers and grown strong vegetation (leaves, stems etc.), repot them in 3 or 5 gallon containers (more notes on potting/repotting to come) before moving them to a dedicated "flowering room" that should be illuminated 12 hours a day by at least one high-pressure-sodium (HPS) HID lamp (most HID setups include a heavy box-like conversion "ballast" and the "lamp," which includes the "hood" and the actual bulb that's to be screwed-into the hood) - this 2 stage rotation of vegetation to flowering is preferred not only due to the considerable difference in the *spectrum* of light favored between vegetating and flowering plants, but because of the very necessary light *period* (the "photoperiod") difference required of the vegetating and flowering marijuana plant - flowering plants are late-summer stage plants that

take advantage of the sun's longer seasonal arc through the sky to soak-in many more daylight hours ("kilowatt-hours" indoors) that do vegetating plants sitting beneath the lower-arcing spring sun - this is *especially* the case when they're pure Sativa or Sativa-dominant strains derived from tropical, equator-hugging regions that enable them to make use of more hours of light per day than their northern and southern-hemisphere-derived Indica-dominant cousins due to the sun's longer arc through the sky in *all* seasons the closer one is to the equator - because pure Sativa's (just as in the animal kingdom, in the plant kingdom *true* pure-breeds are being bred-out and becoming increasingly rare – we're all turning into "mutts," and for the better health of all, as housing a better-rounded, richer blend of hereditary traits makes one, whether plant or animal, more adaptable, less-vulnerable to disease and defect and stronger and more resilient in general) and Sativa-dominant strains can efficiently make use of more light per day, some Sativa growers choose to leave their vegetation lamps on 24/7 instead of the more commonly-recommended 15 to 18 hours per day, with 18 being the maximum number of light-hours that most vegetating plants are generally considered to be able to effectively utilize in any 24 hour period – *flowering* plants, on the other hand, should receive no more than 12 hours of light and an even 12/12 light-to-dark photoperiod every 24 hour day - transitioning from the 18/6 or 16/8 commonly-recommended vegetating period to the 12/12 flowering period by going from, for example, an 18/6 baseline to 16/8 the next week and 14/10 the week after that, all in the vegetation space, and *then* switching to the 12/12 necessary to signal the flowering stage, has been found to be unnecessary at best and counter-productive at worst - while an ideal transition in light-hours per day occurs naturally in the outdoors, it has been found that the plants quickly 'get the clue' that flowering season has begun when the grower institutes a clean break from the 18/6 or 16/8 vegetation-based photoperiod to the 12/12 flowering area's photoperiod, thus rendering needless any such transition as recently exemplified – one of the *many* unsettled matters among experienced indoor marijuana growers is the ideal amount of time plants should spend in each of the two stages of their indoor life cycle - depending upon the particular strain, most of the popular literature asserts that the optimal return-for-investment is achieved with 2 weeks of vegetation, providing clones/seedlings one week to form their first vital roots and a second week to form a decent vegetative base - other growers, myself included, prefer to give their plants a good *month* in the vegetative stage of their development - this one month or so of vegetation also pairs well with those that use or plan to use a clean 2-part rotation with the vegetation area consisting of the first part and the flowering area making-up the 2nd, ideally with the flowering area possessing at least 2 HID lamps, each producing enough light to "finish" 8-15 well-vegetated plants newly-transferred from the vegetation area in about 2 months or so - as soon as one crop of 8-15 flowered females is cut and ready to begin its drying process another set of 8-15 has developed to the point where they're ready to be rotated-in from the vegetation area to fill the vacated space - this setup of 1 dedicated vegetation area and 1 dedicated flowering area is one of the simplest possible examples of a "perpetual harvest system" whose motto might be: *no waste of time, space, energy or money* – once established in the flowering area your plants should show their first signs of beginning to form flowers in 1 to 2 weeks of being changed-over to the 12-hour-on, 12-hour-off daily cycle, as plants interpret this abrupt change in photoperiod as an unmistakable signal of the approaching fall and start to produce the pistils of fertilization and propagation which, if left 'unsexed' by nearby flowering male plants, will produce your desired female crop of cannabanoid-laden sinsemilla (seedless) marijuana plants - if permitted to be fertilized by nearby males, on the other hand, the production of females will switch to focusing on seeds, or propagation, and they'll produce a pitiful harvest of buds by comparison to the unfertilized female of the same strain and general health, both in terms of the quantity and quality of the ripe, finished flowers - keep in mind that this switch to 12/12 is the *only* signal your plants will need that it is time to start flowering, and that this signal will be clearly received *regardless* of the spectrum or intensity of light(s) utilized or whether or not you take my advice of maintaining separate vegetation and flowering areas – please keep in mind that it is absolutely critical that flowering plants receive an *uninterrupted* period of twelve hours of darkness every day, else they could become confused as to the season and are at risk of reverting back to vegetative growth, perverting their life cycle and *grossly* reducing their final yield quality/quantity - if you must inspect your garden area(s) during their 'dark period,' use a green light in your lamp(s)/flashlight(s), as green light reflects off of instead of being absorbed by the green leaves of the plant, thereby *not* interrupting their 'sleep cycle' - make sure plants receive no more light during their 12 daily hours of darkness than would be given-off by the full moon on a clear night (plants evolved to grow *outdoors*) – for reasons that thus become clear it is *highly*-inefficient to utilize the same room or general space for both the vegetative and flowering stages of the marijuana plant's abbreviated indoor life cycle (outside they live for 5 or 6 months if they're planted or their seeds are sown in the spring, versus about 3 months or so indoors), as the vegetating plant can effectively use up to 6 more hours of a different general spectrum and intensity of light *and*, as just mentioned, because maintaining one dedicated area for both the vegetating and flowering stages of the plant's life cycle permits the use of a "perpetual harvest" system of ongoing rotation whereby, once the system is fully established (and depending upon the relative space utilized and the number of plants being grown), the Sonoma County medical marijuana user can legally and conservatively anticipate a yield of ten ounces of dried and cured final product *every month* – regarding the overall growth of marijuana plants, while many assume at first (as I did) that flowering plants will focus *solely* on producing flowers/buds, this is not so – anticipate an additional growth of 1 to 3 feet or so in height and a corresponding increase in both width and foliar density during the 8-12 week flowering period, with pure Sativa and Sativa-dominant strains generally growing taller and wider and taking loner to fully "mature" (due, again, to the longer summers near the equator) in their production of THC and general cannabanoid-containing resin glands, or "trichomes," the source of the marijuana plant's medicine (clearly visible using an inexpensive magnifying glass), which will be most-concentrated on the plant's swelling, ripening buds and their surrounding, tightly-bound leaves (this condensed formation of the biggest and usually most-potent buds and the smaller surrounding leaves that form at the top of the maturing female marijuana plant is sometimes referred to as the "cola" of the plant, or literally "tail" in Spanish due to its cat tail like appearance) – while peak ripeness can be roughly estimated by making note of the progressive maturation of the "pistils" [the hairy follicles growing out from ripening buds that evolved to catch the passing pollen released by any nearby male plant(s) - this captured pollen is then naturally funneled down into the female plant's womb-like "ovule" for fertilization and subsequent propagation-focused seed production] of the flowering female marijuana plant(s) from a vibrant white to a decaying, deep orange color and general appearance corresponding to the plant's exhausting its efforts to reproduce (hence the title of the poem I included at the end of this outline), a much more accurate reading of peak ripeness and cannabanoid potency can be attained using a basic magnifying glass to identify the progression of the trichomes that most casual marijuana smokers refer to simply as "crystals" or "THC crystals" from a translucent white to a milky white and, finally, to a deep-orange or 'amber' color with fat globs (or "heads") of sticky, medicine-packed resin sitting atop - harvest your plants right when the trichomes *start* to turn amber for the best possible cannabanoid profile production out of most strains - <u>note: THC, or "tetrahydrocannabinol," is only the best-known of the cannabanoids that are *together* responsible for producing the desired therapeutic effects in the body and mind of the medicinal marijuana patient</u> - make sure the distribution of light throughout your growing area(s) is as even as possible to avoid imbalances in growth and production, and that the lamp(s) is/are the proper distance from the top of the plants - <u>keep the tops of your plants, collective called the "canopy" of the garden, as close to the light source(s) as possible without putting plant top(s) in peril of overheating, drying-up or literally burning (this distance should generally be no closer than 6" for cooler-running non-HID's and *at least* a foot for HID's) while keeping ever in mind that the intensity and value of light to plants diminishes at an exponential rate with every inch put between your plant(s) and the actual light source(s)</u> - hang lamps from specially-made retractable light hangers so the ideal distance between the lamp(s) and the tops of your plant(s) can constantly be maintained via adjustment corresponding to plant growth/height - fasten a thermometer coupled with a hygrometer (a humidity gauge) above every garden area and hang its sensor down to within close proximity to the top(s) of your tallest plant(s) - avoid allowing the temperature at the top(s) of your tallest plant(s) to climb over eighty degrees or so - remember that tropical Sativa's prefer a bit higher degree of heat and relative humidity than do their colder, dryer-climate-cousin Indica's, so they can be placed closer to light sources - keep moving your plants around (wheels affixed to small wooden flats are convenient for this) so that your shortest plants remain in the middle of your garden and your tallest stay around the outside (think of the ideal shape that might be overlaid atop your garden as being an upside-down dome, assuming you have noticeable taller plants, which won't always be the case) to even-out overall garden growth and keep any already undersized plants from suffering from any further deficiencies - if you have pots to spare, a good trick is to use your old pots from your most recent harvest as height-building-blocks for your shortest plants that can be propped-up and placed in the middle of your garden(s), as you'll likely end-up discarding the used medium(s) contained within those container(s) anyway (assuming it's a compactable growing medium similar to the coconut coir I use) and since the used medium(s) within those container(s) will conveniently collect the excess water and nutrients that drain from the container(s) in which the plants are actually contained – <u>regarding</u>

drying: cut your finished females (again, to be judged by their resin gland and pistil maturation) at their base and hang them with string or wire etc. in a cool, dry place (i.e. *not* in the same room or area that you grow - the humidity in grow areas will delay drying at best and lead to rot and a partial or total loss of your crop at the worst) - wait for the stems to stop flexing and start cracking or breaking when bent as a sign that your plants are dehydrated enough to 'trim' off the buds - regarding trimming: it is advised that you detach and manicure all of your crystal-laden buds over a 'kief-catching' trim tray that you've acquired at a hydroponics supply store or that you've made yourself with a fine mesh silk screen etc. so that you catch some of the "kief" (= the trichomes/resin glands that no longer shimmer like crystals because they're fully dehydrated) immediately and don't waste the considerable amount covering the best leaves (large 'fan leaves' produce poor concentrates because they contain relatively few trichomes - the closer the leaves grew to the main cola of the plant the richer in resinous trichomes they'll be) - you can simply smoke all of your collected kief or, after you've trimmed all of your finished females over trim tray(s), you can freeze that trim (because frozen resin glands break-off easier and are thus easier to separate from the plant matter) and, later, with a little cold-water extraction and pressurized heat, process it into grade-A "bubble hash!" (the purest hash bubbles and leaves a whitish residue when burned - kief is much easier to handle, store and smoke when congealed and pressed into 'hash balls' - Cervantes recommends the book *Hashish!* by Robert Clarke and the site *www.pollinator.nl* for more on making hash and other cannabanoid concentrates from your medical marijuana harvest) - a warning regarding kief and its derivative cannabanoid concentrates: these dried, pulverized resin glands have a much higher cannabanoid content than marijuana buds by themselves because these dried resin glands are the actual source of all the medicine and, if well-processed, should contain a minimum of plant matter, *so be careful when smoking or consuming it* - regarding curing: once you've trimmed your plants allow the trimmed buds to finish drying in your trim tray(s), in open bin(s) or in other breathable container(s) for a week or so, at which point they will generally be ready to be smoked and can be stored for "curing" in sealable jars or even paper bags - the 2-part step-down from paper bags to sealable jars such as masons jars makes for an excellent 2 week drying process where you switch to the sealable jars after the 1st week or so for the rest of the curing and general storing and potency-preservation process - remember to open-up your curing container(s), regardless of their make, at least once, preferably twice or more a day to allow any remaining moisture in your *almost* completely dried-out buds to escape, as not doing so could result in the remaining trapped traces of moisture leading to fungi or rot and literally spoiling all your hard work - HID lamps, which are usually metal halides (MD's) during vegetation and high-pressure-sodium's (HPS's) during the 8-12 weeks of flowering depending, again, on whether the plant(s) are an Indica or Sativa-dominant strain, produce *much* more heat than the CFL or LED lamp(s) often used during vegetation but generally *not* recommended for flowering as neither technology is yet to consistently demonstrate the consistent production of intense-enough light for full-potential flowering - HID lamps need to be placed further away from the tops of plants than do CFL lamps or LED lamps - these more efficient compact-fluorescent bulbs and lamps and, *especially*, the relative-newcomer LED lamps, which remain too expensive and generally not worth the cost-to-benefit ratio *at this point* in their evolution and application to indoor horticulture but are *far* superior in their potential technological capacity to convert electricity into light at a minimum of energy wasted to heat when compared to the now relatively-outdated HID technologies - <u>once LED ("light-emitting-diode") technology's raw materials and manufacturing processes are better-perfected and the corresponding price for LED equipment drops to affordable rates, LED lamps will take their place as the preferred lighting technology of proficient indoor medical marijuana growers around the world</u> - while both CFL's and LED's currently produce significantly fewer lumens-per-watt than do HID's and therefore generally can't provide the same overall illumination value to "full sun" plants like marijuana, these more up-to-date illumination technologies also lose *much* less energy to heat for every watt they pull from your wall socket(s), and heat production is not only very expensive but can choke and even cook plants while making the entire garden more difficult to manage - <u>in terms of the angle(s) at which you should supply the light from your lamp(s) to your garden(s), make sure to hang your lamp(s) directly above your plants in both the vegetation area and the flowering area</u> - sideways-supplied or "horizontal lighting" is inefficient and not recommended, as a considerable amount of the light being rendered isn't actually being absorbed by the plants' leaves, but instead passes or is 'filtered-through,' as leaves, while naturally adaptable to the angle of light to which they're consistently supplied [you'll distinctly see parts of the entire plant grow and 'reach' towards the closest and/or strongest light source(s)], generally grow to take in light from the 'sun' above - this is one of the reasons that Sativa-dominant and especially pure Sativa strains are considered less suitable for indoor growing, as their much more substantial height and girth makes providing light to their 2-3 feet or more of low-lying leaves very difficult without side-lighting which, again, generally isn't worth the expense - the natural equator-hugging ecosystems of the Sativa plant receive not only more intense light and more light hours per day than do those natural environments of the Indica plant but also a greater average *distribution* of light along all sides of the developing plant - it is also recommended that at least the flowering room's light(s) be run during the night when suspicious *PG&E* meter-readers that might feel the need to report the excessive use of electricity during the daytime won't be around to do so - I run my 1000W HPS flowering room lamp from 7pm to 7am and my 3 125W CFL vegetation lamps from 1am to 7pm - running your HID lights at night instead of during the day has the added benefit of allowing you to better protect your flowering females from overheating and excessive humidity during the warmer months of the year - with the use of electricity in residential areas in general, always know how many amps each area of your residence can draw at any one time before they overdraw and 'trip' the circuit breakers, temporarily killing your grow project, which can be a serious problem if you're not around for extended periods - avoid bothers related to your intense grow lamp(s) triggering unwanted attention, "black-out" windows with specially made or makeshift materials in such a way that doesn't too-severely restrict your windows' capacity to draw-in the fresh air and/or, depending upon the particular windows' strategic use (see #3 herein), force-out the used air, both elements of which are vital to the good health and robust growth of your gardens - <u>please note that *none* of this is meant to be an implicit approval of illegal</u>

marijuana grow projects, but rather a recognition of the potential issues related to the medical marijuana grower attracting too much attention from potentially suspicious police, residential complex managers and prejudicial neighbors with whom they must coexist and, thus, whom they should be careful not to provoke - *critical cost savings note*: most residential districts are on graduated cost scales enforced by *PG&E* such that the more electricity you use the more likely you will qualify for their next graduated cost-per-kilowatt-hour-used scale, meaning you'll thenceforth pay more for *all* the electricity you utilize, with each level increasing *substantially* in cost per kilowatt-hour for every additional kilowatt-hour utilized over the 'cap' of the previous scale - in my area, for example, there's a 'baseline,' or starting scale, of just under 13 cents per kilowatt-hour charged for the first 351 hours, then almost 15 cents for the next 105 hours, *30 cents* for the 245 hours after that and, finally, *34 cents* for every "Kwh" I utilize on scale #4 - *with your costs per kilowatt-hour more than doubling as you reach the 3rd scale*, it is recommended that homeowners utilize solar, wind etc. power to supplement the electricity they draw from *PG&E* and concordantly limit their electrical costs - one last note regarding the use of HID hoods with glass enclosures separating the bulb from the plants: it is generally recommended that you remove this, as a significant degree of the light produced by the bulb is reflected off of this glass surface instead of passing through to your garden, thus hampering ideal growth - just keep in mind that this will release more heat onto your garden as well, so supplement with circulating fans and/or add the right-sized inline fan to the hood

5. **Manicure your plants within 10 days of moving them to the flowering area** - of all of the lessons related herein, this is the one that I actually learned last, and it is of significant importance for the at least the following catalogue of reasons:

   a) Your female plants will produce fewer, denser and larger, easier to market buds due to the horticultural property of the nutrients, water and energy of plants being distributed up the main stalk of the plant and out to each of its branches; therefore if you manicure your plants before they are too far into producing their flowers that total productive capacity is diverted to fewer flowering or budding points on the plant, making each bud of the plant not only bigger and denser but usually more potent as well
   b) Due to the above (a), you won't have to worry about trimming and marketing overly-leafy and less-potent lower buds that don't develop properly in indoor grow areas due to the fact that there is usually no arc in the daily providing of artificial sunlight produced by your grow light(s) (unless you utilize automated light tracks and other expensive equipment) and, thus, the lower portions of flowering female marijuana plants will receive insufficient light from the proximate leaves to produce dense nugs
   c) Related to the above (b), less, bigger, denser nugs means a much easier, less time-intensive and money-consuming trimming process, which can take *much* longer than people realize (it takes me about 2 hours to fully trim a larger harvested Sativa female cannabis plant), saving you time, money, energy and the beginnings of the onset of carpel-tunnel-syndrome
   d) Your flowering female marijuana plants will, due to their being less super-leafy condensed formations at the lower levels of the plant, receive much better air-circulation and benefit significantly from the related increase in oxygen and CO_2 uptake levels - all of the points under this

heading are, of course, interrelated (*everything* is connected in science as in life), but this particular point is worth stressing more than most, in my opinion, because an open lower portion of the plant that permits easy pass-through of fan-produced artificial winds and nutrient deliver also means…
e) <u>Your flowering female marijuana plants will be much less prone to excess moisture being trapped close to the plant and, thus, will be much less likely to suffer the serious risks associated with an excessively humid environment such as drowning roots and especially fungal colonization's</u> of, as one example, the aforementioned 'powdery mildew' that invaded my 1st crop of White Widow plants; this can save entire cannabis crops
f) <u>Assuming you employ the so-called "Drain-to-Waste" watering-by-hand method that I employ to water and feed your flowering female marijuana plants, you will find watering/feeding plants is *much* easier without the lower branches getting in the way</u>; this is especially true with regards to those plant(s) that sit in the middle of your garden surrounded by other females on all sides; without cutting-away the lower branches the hand-watering of these center plants is much more difficult than necessary
g) <u>Having significantly fewer branches overall means your flowering female marijuana plants and the pots that house them will be considerably lighter, which means they're easier to move-around as relative garden growth dictates and for general maintenance and garden control purposes</u>
h) Last but definitely not least among the positive benefits to be garnered from the manicuring of your flowering female marijuana plants is that cutting-away the lower branches (or at least cutting-away any secondary branches branching away from any core branches 'branching-off' of the main stem or 'stalk' of the plant) means cutting-away the branches that add girth to the plants by reaching-out horizontally for the usually top-down 'vertical lighting' but that don't significantly add to production, which means <u>the total garden and canopy space is reduced without reducing productivity, which further means that less total space is being consumed per plant, that productivity per square foot of garden area is significantly increased and that, also of considerable value, there is less need of leaving room for the outer perimeter mentioned earlier in this outline used for watering and general maintenance of your garden(s)!</u>

6. **Always <u>have a plan, take notes and track results – assuming you want to Improve</u>** - be a "man (or woman) with a plan" - the difference between planning and execution reveals invaluable information every time - regardless of whether or not you hit your mark(s), it is *always* worth it to have some idea of what you're aiming for in order to eventually be able to hit your target at will - determining *why* your results differed from your projections is the first step both in building more accurate projections and in identifying and improving-upon those elements of your project that, upon reflective analysis, left you short of your goals - <u>the following example illustrating the current state of my own amateur grow project is meant to demonstrate how the accounting of a medical marijuana horticultural project might look in my area of Santa Rosa, CA – first, know your legal limits…</u>

SONOMA COUNTY, CA's, MEDICAL MARIJUANA LEGAL GUIDELINES PER LICENSEE:

- *NO MORE THAN 100 square feet of total growing space (including canopy)*
- *NO MORE THAN 30 plants grown within that 100 square feet of total space*
- *NO MORE THAN 3 total pounds in your personal possession at any one time*
- *ALWAYS POST COPIES OF YOUR UPDATED DOCTOR'S RECOMMENDATION NEAR ANY AND ALL MEDICAL MARIJUANA GARDEN AREA(S) YOU MANAGE*

ESTIMATIONS FOR PRODUCTION, EXPENSES, AND PROJECT BREAK-EVEN POINT –

Note: All following estimations assume a *combined* cost or value of sold *and* consumed product

EXPENSES:

1) **With my current set-up** of vegetation closet and one 1000W flowering lamp: approximately $600 for flowering room rent expense, $100 for electricity, $20 for water/sewer/trash and $50 for nutrients and dips (*not* counting the cost of clones or equipment depreciation) equals an **estimated monthly expense of $770.**

2) **With my planned set-up** of vegetation closet and *two* 1000W flowering lamps: approximately $600 for flowering room rent expense, $200 for electricity, $30 for water/sewer/trash and $90 for nutrients and dips (not counting the cost of new clones *or* the recouping of equipment expense, which will increase with both the second HID lamp/bulb as well as with the planned acquisition of a carbon filter and two or three inline fans) equals an **estimated monthly expense of $940.** With the fixed cost to use the apartment space for the project it makes *much* more fiscal sense to run *two* flowering lamps in that 2nd bedroom (the "flowering room") so that the rotation of approximately 4 weeks or so in the vegetation closet and 8 to 12 weeks to flower permits the harvesting of ten to fifteen plants per month on avg. versus only being able to harvest half that often and wasting the closet space when using only one flowering lamp in the flowering room; there's a **drawback in conspicuousness with this setup if *PG&E* decides to report your use, so *always* pay your bill on time. NOTE: With *this* planned example of a perpetual harvest indoor medical marijuana project the monthly expenses, *NOT* including clone cost or the $600 I've included for the renting of the required extra room, can be broken-down to around 9 cents per watt per month, including about 7 cents per watt per mo. for the electricity itself and another 2 cents per/W/mo. for all other project expenses (water, trash, sewer, nutrients, preventative sprays/bathing mixtures etc.)**

PRODUCTION:

1) **With my current set-up** of vegetation closet and 1 1000W flowering lamp: about ten plants can be harvested every *two* months or so, making the total *monthly* production an average of five plants per month; with a semi-conservative estimated one ounce per plant you get five ounces per month (though, again, you also have to *wait* for two months in this setup, so you're not *really* getting five ounces per month but ten every other month) which, if sold for about $180 per ounce on average equals an **estimated monthly income of about $900, putting the current set-up's overall monthly value at about $130. With production of 1.5 O's per plant on avg. (7.5 O's/mo.), by comparison, the set-up's worth $580/mo.**

2) **With my planned set-up** of vegetation closet and <u>2</u> 1000W flowering lamps: about ten plants can be harvested every month or so; with a semi-conservative estimated one ounce per plant you get ten ounces per month which, of sold for about $180 per ounce on average equals an **estimated monthly income of $1,800,** putting the planned set-up's **overall monthly value at about $900 though, again, *this is a conservative production estimate* in both production per plant and value per ounce.** One and a half ounces per plant is very much possible, and even two+ O's per plant is attainable. ***At 1.5 O's per harvested plant you get about one pound per month under this setup which, at the same average of $180 per ounce, would yield a gross income of about $2,900 per month, or an overall (net) monthly value of about $1,660, which would make the grow project WELL worth the costs.*** Though I've yet to turn out anything near it myself, *I've heard of big-producing strains like "Blue Dream" and the classic "White Widow" strain generating <u>5 to 7 ounces per plant</u>, sometimes more, even indoors! But even if you reached the level where you were averaging 2.5 O's per plant your gross income value would be about $4,500/month, putting your project value at about $3,550/mo.!* Despite the potentially-legendary production of the best growers, beginners and amateurs should, while of course always pushing and hoping for more, shoot for production of 1.5 O's per plant. <u>*Cervantes and other experienced growers suggest shooting for about twice that, or about one gram of final product per watt of lamp(s) used.*</u> With the fixed cost to use the apartment for the project it is almost necessary that 2 lamps be used in the flowering room so that the rotation of about 4 weeks or so in the vegetation closet and 8-12 weeks to flower allows the harvesting of 10-15 plants/month, making the project far more attractive than with 1 flowering lamp.

BREAK EVEN POINT:

1) **With my current set-up** of vegetation closet and **1 1000W flowering lamp** the total estimated monthly expenses related to the project are about $770 which, broken-down to a per plant basis and assuming 5 plants are harvested (half as many as under planned setup), produces a **cost per harvested plant of around $155.** To break even, the final product has to be sold to legal dispensaries or valued as personal medicine at $155 per the harvest of every matured, cut, trimmed and cured plant. **Anything brought-in or valued at over $155 per plant may be considered "in the black (a profit)," anything less than $155 per plant may be considered "in the red (a loss)." *At a cost of almost $65 more per plant than with the planned setup, my current setup is highly-wasteful.*** On a per *ounce* basis using a 1.5 ounces per plant average (7.5 ounces per month on average under this setup), the amount at which every ounce must be valued or sold to a dispensary for the grower(s) to break-even on their comparable project is **around $100/O.**

2) **With my planned set-up** of vegetation closet and <u>2</u> **1000W flowering lamps** the total estimated monthly expenses related to the project are about $940 which, broken-down to a per plant basis and assuming 10 plants are harvested (twice as many as with current setup), produces a **cost per harvested plant of around $94.** To break even, the final product has to be sold to legal dispensaries or valued as personal medicine at $94 per the harvest of every matured, cut, trimmed and cured plant. **Anything brought-in or valued at over $94 per plant may be considered "in the black (a profit)," anything less than $94 per plant may be**

considered **"in the red (a loss)."** On a per *ounce* basis using a 1.5 ounces per plant average (15 ounces per mo. on avg. under this setup), the amount at which every ounce must be valued or sold to a dispensary for the grower(s) to break-even on their comparable project is **around $60/O!** *Closer to the best possible scenario, if 2.5 O's per plant were managed under this setup (25 O's/mo.), ea. O would need to be valued at just $36 for the grower(s) to break even.* **Therefore, once high quality buds are being produced the grower(s) will likely be happy with this setup.**

Final Note: You get-*out* what you put-*in*, and not just in terms of cultivating marijuana

Author's Acknowledgements and Disclaimers

Most of the information contained herein was adapted from 4 sources: online research, conversations with friends and fellow aspiring cultivators, the experiences of the grower-author and what I recall from the *much* more considerable experiences of others related through valuable literary resources such as *High Times Magazine* and *especially* Jorge Cervantes' acclaimed horticultural guide: *The Medical Growers Bible*, essential reading for anyone that hopes to someday become adept at the science of cultivating marijuana.

This document is meant for the informational purposes of licensed medicinal marijuana users and growers and for indoor horticulturalists in general, to whom the majority of the information and "lessons" will apply. While every writer hopes for success, I wrote this document not just for myself but out of a desire to assist *legally* growing and consuming marijuana enthusiasts around the world that have discovered the medicinal and, I would argue, *the general soothing, appreciation-of-life-enhancing offerings of the miraculous marijuana plant.* Therefore I ask that you please not consider this document a reason to accuse me of inspiring or endorsing illegal marijuana growing operations or of being the cause for the failure of a harvest because, say, you think I omitted some particular bit of vital information. This document does *not* have all the answers - it is, like in *The Pirates of the Caribbean*, "more like a guide than a rulebook," and should be treated as such.

Finally a 1st draft of an apropos poem I'll include in an upcoming collection entitled "*Love of Wisdom:*"

Sexually-Frustrated Females

Beam me up Scotty
'Cause I'm ready to explore space
Ready to spread the blessings of Romulan
On a mission of foremost urgency and haste
Knowing how much less blood might be shed
With aliens rolling green instead of seeing red

After preventing enough intergalactic dueling
And feeling my spaceship grumble for refueling
I'll return to my own conflicted terrestrial base
In order to spread this secret shield of peace
To every frustrated member of the human race

A garland of ganja I'll wrap around their necks
Topping their pipes with dried cannabanoid flecks
Off my prized colas I'll break them a generous piece
For the far more peaceful lives they now stand to lease

Standing starkly in contrast to those drowning in booze
Living pathetic, frightened lives with nothing to lose
Cretans filled with hatred and fear - the type to avoid
Yes, armed with guns and big, awful blades they may be
Angrily firing lasers out their eyes at everything they see
In their confused, frenzied minds they're anything but free
Everyone's out to get them, and they call us paranoid
Yet I want only to help them to feel less annoyed
For no drug transcends hate like the cannabanoid

Give me that spicy Kush for when it's time to chill
As I'm not one to swallow the 'ol 'unwind-in-a-pill'
Purples are a fruitier way to calm yourself down
And turn that sad little frown entirely upside down
While the Whites will leave you flying way up high
Floating easily along through the shimmering sky
Or Poison yourself with that Rift Valley Durban
The best African connection brought into the urban

In my garden I bring meditation to life
Mary's wearing my ring; already my wife
Fruit for a mental change of perspective
For your pain and hostility, a potent corrective
Ripening flowers covering a crystallized plant
Even with the Feds still claiming you can't
No chemical fillers to this product added
Nothing cut by the knife, coated or padded

No disadvantaged, exploited youths contained within
No AK47's carried to cover my grey market goods
So approach without dread of some forthcoming sin
Without crossing lines marking where felons have been
Guns, blades, toxins and explosives are nowhere near
Walk down my upscale walkway with far less to fear
Than approaching thugs packing heat in the woods
Hiding cartel plots shrouded in menacing hoods
And suddenly risking everything you hold dear

Like the scientist, experimentation is essential

To become the cultivator tokers treat reverential
For every strain has something new to teach
A new mental state of being to reach
A new taste, high and perspective to add
Another way to realize the blessings you've had

Do you possess the cure to everything that ails me?
To all that stresses, straightens and bores me?
A way to forget about 'the man' that whores me?
Or about the peace that eludes and deplores me?
For after the very necessary daily work is done
And I'm ready to kick-back and have some fun
To my jar of cured sinsemilla flowers is where I run

Resinating with medicine for the body and mind
Mary's gifts of good health are one of a kind
But not just for the sick, the restless and weary
Her greatest gift may be in deconstructing the dreary
Until you feel the force and gravity of every moment
Your imagination now bursting at the seams
With vivid new pictures your mind reels and teems
The once trickling stream now begins to foment

Smells, colors, hungers and passions awaken
Look what priceless endowments were once forsaken
It's as if you've been taken softly by the hand
By the most benevolent goddess in all of the land
And lead down a path to a pristine, parallel place
In an equananimous spirit, at an unhurried pace
Through doors shimmering with crystalline trichomes
To a place where splendor and rapaciousness roams

Where the appreciation of everything gifted
Is assured before it can even be wrapped
Where the experience of every sensation
Is set for amplification before it can be capped
Where the greatest degrees of pain and pleasure
Are like bottomless reserves never to be tapped
And the greatest insights of the heart and mind
Are presciently outlined before they're mapped

To every fearful person that moves to pick-up a gun
In order to tear into the flesh of his fellow man
Due to disagreements and religious speculations
On the 'one way' to our spiritual emancipations
And the name of the Spirit or how it all began
Must first pray at the alter of the Sticky Green Leaf
And smoke sensei sin similla whenever they can